Happy Veggies™
Happy Garden™ Series
Storybook

Kid Friendly String Bean Casserole Recipe Included

Sidney String Bean

With A Little Help From My Friends!

J Stephen Sadler

Illustrations by Hatice Bayramoglu

Happy Veggies ™

Happy Garden^{TM} Series

Sidney String Bean 8

Copyright © 2021 by J Stephen Sadler

All Rights Reserved. No part of this book may be reproduced or transmitted in any form or by any means, electronic or mechanical, including photocopying, recording or by any information storage and retrieval system without permission in writing from the author.

ISBN: 978-1-953578-12-9
Library of Congress: 2021905129

Sadler Media, LLC

Printed in the U.S.A.

www.JStephensGarden.com

DEDICATION

To my grandchildren Charley, Jude, Liv and Evelyn who help me everyday see the wonders of life through a child's eyes.

It was Spring time and Farmer Bob, as he did every year, was busy planting his garden with the best vegetables in the world. First he planted his tomatoes with their special cages to keep them from falling over.

Next came his little Eggplant plants with wire rounds to hold up their soon to be BIG and PLUMP eggplants.

And then on to his beautiful looking baby spinach and of course his cauliflower plants placed ever so carefully in the center of his garden to keep Colinda Cauliflower from venturing to far away.

Next were his carrots and potatoes in the front of the garden because everyone knows they grow down into the ground and not up.

Finally Farmer Bob planted his string beans with a BIG black wire cage so they can grow up, up and up.

There were rows of string bean vines each with tiny string beans. Much smaller and weaker than all the other baby string beans, Sidney started life barely holding on to the string bean vine. Can you find little Sidney?

As they reached for the sun and drank from the Spring rains, all the other baby string beans grew stronger and plumper but, poor Sidney withered under the shade of their leaves.

Farmer Bob did his best to try and help little Sidney, moving the leaves of the other string beans to provide more sunlight. But, because string beans all share each vine, Farmer Bob couldn't help little Sidney with more fertilizer or water.

Farmer Bob was sad as he checked on Sidney every day but, he knew there was nothing more he could do to help his poor little string bean. It looked like it was only a matter of time until little Sidney String Bean would wither away and die.

But Sidneys' friends would NOT let THAT happen. One night while little Sidney was fast asleep, they made a pact to all work together to save their little friend.

Stella and Sherman string beans who were on each side of little Sidney said "We'll move our leaves to make sure little Sidney gets the most sunlight!"

Stevie String bean volunteered to share some of the water he gets with little Sidney and Selena String bean said "I'll share some of the fertilizer that comes my way!"

All the other string beans on the vine agreed that they wanted to share all the water, sunlight and fertilizer that they received to help little Sidney grow big and strong.

And from that day on, all of little Sidney's friends did everything they could to help little Sidney get big and strong.

Right away, Farmer Bob noticed a change in his smallest string bean. At first Sidney's green color started to look like all the others.

As the month's went on, Farmer Bob noticed that little Sidney was no longer LITTLE! He was becoming BIGGER and BIGGER and all his friends were SO proud of how their little friend had now become the BIGGEST string bean of the bunch.

One day, Chef J Stephen came to the garden to pick the BIGGEST and the BEST Veggies for his special bistro dishes. When he came to the string bean patch he saw the BIGGEST and BEST string bean he had EVER seen! And THAT was..... "LITTLE" Sidney!

When Chef J Stephen placed no longer "LITTLE" Sidney into his basket, "LITTLE" Sidney jumped up to the Chef and whispered in his ear "I may be the biggest string bean but the smaller string beans are ALWAYS the sweetest. You should select them and make a special string bean casserole dish that will become the most famous string bean dish in the world!"

Chef J Stephen LOVED Sidneys idea and selected ALL of Sidney's friends, even the smallest of the lot to make the most famous string bean dish in the world.

You see, "LITTLE" Sidney, who wasn't little anymore, remembered what his best friends did for him when he was struggling to stay alive and he wanted to do SOMETHING special for THEM.

"LITTLE" Sidney had learned that doing good things for others is a good thing for everyone!

- THE END -

Happy Garden Series

J Stephen's "Happy Veggies" books are designed to let kids know that there are vegetable dishes that even they will like. Healthy eating explained in a way that kids will enjoy is at the core of every one of his "Happy Veggies" children's books.

We understand that once your children read Sidney String Bean, they're going to want to try their very own Sidney String Bean casserole recipe. Not a problem. Chef J Stephen has made sure that you can make your own "healthy" Sidney String Bean casserole recipe by providing you with his own recipe for you to make. It's easy to do and a great project to do with your children.

We hope your children enjoy the book AND the Sidney String Bean casserole recipe and learn that even vegetables can be, not only healthy, but delicious as well.

Yummy in the Tummy Sidney's String Bean Casserole Recipe

Ingredients
- 1 can of 10 1/2 oz Campbells® Cream of Mushroom Soup
- 3/4 cup whole milk
- 1/8 tsp black pepper
- 4 cups cooked cut string beans
- 1 1/3 cups FRENCH'S® Crispy Fried Onions

Instructions
- Mix your soup, milk, and black pepper in a 1 1/2 quart baking dish
- Stir in your cooked string beans
- Add 2/3 cup of Crispy Fried Onions
- Bake at 350° F for 30 minutes or until hot
- Stir
- Top with your remaining Crispy Fried Onions
- Bake for 5 minutes or until the Onions are crispy golden

Author J Stephen Sadler

Author, Speaker, Chef, Restauranteur and Travel Host, J Stephen Sadler travels the world to find unique and healthy dishes to bring back to small town America.

His *Happy Veggies Happy Garden Series* of children's books are written to encourage healthy eating in young readers. Each book also features a healthy child friendly, easy to prepare recipe for parents to create for their children.

You can find all of Chef J Stephen Sadler's *Happy Veggies Happy Garden Series* of books at www.JStephensGarden.com, Amazon, Ingram, Apple, other online book platforms and local retail outlets.

Chef J Stephen also speaks to children's groups, schools and organizations about healthy eating. His seminars and discussions are just as engaging for children as are his books. To reach J Stephen to discuss a speaking engagement contact him at: jstephen@sadler.media.

Happy Veggies

Happy Garden Series

Be sure to get all the Happy Veggies storybooks and coloring books in the J Stephen Sadler *Happy Garden Series:*

Colinda Cauliflower Book 1
Bubba Broccoli Book 2
Chuckie Carrot Book 3
Eddie Eggplant Book 4
Sammy Spinach Book 5
Tanya Tomato Book 6
Paulina Potato Book 7
Sidney String Bean Book 8

Happy Veggies single storybooks, single coloring books, box sets of storybooks, box set of coloring books, or the all inclusive Activity Set of storybooks, coloring books, and crayons combined, are available at: www.JStephensGarden.com, Amazon, Ingram, Apple, other online book platforms and local retail outlets.

www.ingramcontent.com/pod-product-compliance
Lightning Source LLC
Chambersburg PA
CBHW061105070526
44579CB00011B/144